Energy 237

太阳能利用的未来

More From The Sun

Gunter Pauli

[比] 冈特·鲍利 著

[哥伦] 凯瑟琳娜·巴赫 绘

贾龙智子 译

上海远东出版社

丛书编委会

主　任：贾　峰

副主任：何家振　闫世东　郑立明

委　员：李原原　祝真旭　牛玲娟　梁雅丽　任泽林

　　　　王　岢　陈　卫　郑循如　吴建民　彭　勇

　　　　王梦雨　戴　虹　靳增江　孟　蝶　崔晓晓

特别感谢以下热心人士对童书工作的支持：

匡志强　方　芳　宋小华　解　东　厉　云　李　婧

刘　丹　熊彩虹　罗淑怡　旷　婉　杨　荣　刘学振

何圣霖　王必斗　潘林平　熊志强　廖清州　谭燕宁

王　征　白　纯　张林霞　寿颖慧　罗　佳　傅　俊

胡海朋　白永喆　韦小宏　李　杰　欧　亮

目录

Contents

一位老爷爷正抬头仰望着自家屋顶上的太阳能电池板。他的孙女也抬起头，说道：

　　"爷爷，您当初在屋顶上安装太阳能电池板可真是太有远见啦。我们可以利用阳光为一切提供能源——从我们的手机到电灯，还能加热我们的水。"

A grandfather is looking up at all the solar panels on his roof. His granddaughter, looking up too, says:
"You were such a visionary, Grandpa, to have put solar panels on the roof. We can power everything with the sun, from our phones to our lights, to warming our water."

......看着太阳能电池板......

... Looking at the solar panels ...

这个村子里第一批竖起的太阳能电池板……

first panels put up in this village ...

6

"是的，你要知道，这些可是这个村子里第一批竖起的太阳能电池板。"

"我为您能这么做感到骄傲，但真希望我能永远为您骄傲……"

"你是什么意思？有什么我本来能做到的事吗？还是我已经做错了什么吗？"

"Yes, these were the first panels ever put up in this village, you know."

"I am proud of you for that, but do hope I can always remain proud of you…"

"What do you mean? Is there something more I could have done? Or something I have done wrong?"

"不，不！您没做错任何事，相信我。只是有些事情是没人能预见到的……"

"你是说你将来不会为你的老爷爷骄傲吗？"

"No, no! You are a saint, believe me. It is just that there are some things no one could have foreseen…"

"Are you telling me you will not be proud of your old grandad in the future?"

你是说……

Are you telling me ...

我所有的邻居现在都安装了太阳能电池板……

All my neighbours now have solar panels ...

"爷爷，我不是在预言什么。我只希望旧的太阳能电池板除了作为我们屋顶上的装饰品以外，还能有别的作用。"

"哦，因为我的所作所为，我所有的邻居现在都在他们自己的屋顶上安装了太阳能电池板。"

"I am not predicting anything, Grandpa. I just want old solar power panels to become much more than just an adornment on our roof."

"Well, because of what I did, all my neighbours now have solar panels on their roofs as well."

"太阳将在未来数十亿年内提供能量。只是第一个在太阳能发电方面迈出一大步还不够。我们不能再考虑得长远一些吗？"

"好吧，我年纪大了，可能没有时间再当开拓者了。"

"The sun will be giving power for billions of years to come. Being the first to have taken that great step forward towards solar power, is just not enough. Can't we think beyond that?"

"Well, I am getting older, and may not have the time to be a pioneer again."

……在未来提供能量……

... power for years to come ...

......还能追求新的梦想吗？

... still chase some new dreams?

"您花了更多的钱来获得太阳能，因为您相信这很重要。但是爷爷，我想说明一件事：您现在就满足于已取得的成就也太早了！"

"那么，你觉得像我这样的老人家还能追求新的梦想吗？"

"我们应该一起追梦！告诉我，一个太阳能电池板能用多久？"

"要我说，估计至少三十年吧。"

"You paid more to have solar power, because you believed it was important. But Grandpa, I want to make one thing clear: You are far too young to be content with what you have achieved!"

"So, you think an old man like me could still chase some new dreams?"

"We should dream together! Tell me something, how long will a panel last?"

"At least thirty years, I would say."

"谁来回收它们？"

"有些公司会拆除框架，分离玻璃和塑料，提取金属，分离铅和镉……"

"非洲有多少妇产医院能用上电？"

"And who will recycle them?"

"There are companies that will remove the frame, separate the glass, and the plastics, and extract the metals, separating the lead and cadmium…"

"And how many maternity hospitals are there in Africa that have power?"

分离玻璃和塑料……

separate the glassand the plastics ...

为世界带来可再生能源……

Renewable energy to the world ...

"回收旧太阳能电池板与另一个大洲的产妇接生有什么关系？"

"您难道不希望太阳能发电成为一个伟大的产业，一个为世界带来可再生能源的产业吗？"

"我当然希望。我知道新技术将会涌现，并使之变得更好。"

"What has recycling old solar panels to do with mothers delivering babies on another continent?"

"Don't you want solar power to become a great industry, one that brings renewable energy to the world?"

"Of course I do. And I know that new technologies will emerge, and make it even better."

"如果把所有即将达到使用周期尽头的太阳能电池板回收起来，在由于这些面板而能用上电的医院里面，每个刚出生的婴儿都能够接受医疗救助，而当您采取必要措施来实现这一目标时……"

"我肯定会成为你永远的英雄。"

"没错，爷爷！"

……这仅仅是开始！……

"When every panel reaching the end of its life is recycled, and every baby starting its life receives medical help in a hospital that has power because of those panels, and when you do what is needed to make this happen…"

"I will deserve to be your hero – forever."

"Exactly, Grandpa!"

… AND IT HAS ONLY JUST BEGUN!…

……这仅仅是开始！……

... AND IT HAS ONLY JUST BEGUN! ...

The Sun blankets the Earth with enough photons every hour to meet the entire world's energy needs for one year. The sun releases energy in the form of light and radiation.

太阳每小时覆盖在地球上的光子足够满足全世界一年的能源需求。太阳以光和辐射的形式释放能量。

The International Renewable Energy Agency (IRENA) estimated, in 2016, that there were 250,000 metric tons of solar panel waste in the world. This amount could reach 78 million metric tons by 2050.

国际可再生能源署（IRENA）估计，2016年，世界上有25万吨的太阳能电池板废料。到2050年，这一数字可能达到7 800万吨。

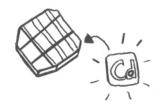

每100万块晶体硅太阳能板中含有30吨镉。每安装一千瓦的发电量，大约有4克铅和23克镉。

There are 30 tons of cadmium contained in one million crystalline silicon panels. For each kilowatt of generation power installed, there are approximately 4 grams of lead and 23 grams of cadmium.

镉是一种存在于地壳中的银白色金属，是铜、铅和锌开采的副产品，在燃烧矿物燃料（煤）、吸烟和随意焚烧废弃物时会排放。

Cadmium is a silver-white metal present in the Earth's crust, a by-product of copper, lead and zinc mining, and emitted when burning fossil fuels (coal), by smoking cigarettes, and by uncontrolled burning of waste.

Since sunlight is diluted and diffused, it requires large collectors to capture and convert the sun's rays to electricity. Cadmium Telluride is the most common thin-film solar cell, but it is toxic to aquatic life.

由于阳光被稀释和扩散，需要大型的收集器来捕捉太阳光并将其转化为电能。碲化镉是最常见的薄膜太阳能电池，但对水生生物有毒。

Solar power needs 16,000 tons of materials for one terawatt hour of power. Hydropower requires 14,000, wind power 10,000. Most efficient is geothermal power (5,000), and kite power, with a mere 1,000 per terawatt hour.

太阳能发电每产生1太瓦时电能需要消耗1.6万吨原料。水电需要1.4万吨，风电需要1万吨。效率最高的是地热能（5 000吨）和风筝发电，后者每太瓦时只需消耗1 000吨原料。

非洲一年中接受的阳光比任何大洲都多。摩洛哥的努尔是非洲最大的太阳能项目所在地。南非拥有非洲大陆十大太阳能发电厂中的八座。

Africa receives more sunshine during the course of a year than any continent. Noor, in Morocco, is home to Africa's largest solar project. South Africa hosts eight of the ten largest solar plants on the African Continent.

东撒哈拉和非洲东北部保持着日照的世界纪录，每年可达4 300小时（或达到潜力的97%）。非洲6 000万太瓦时/年的发电能力是全球太阳能发电潜力的40%。

Eastern Sahara and North-Eastern Africa hold the world record for sunshine, with 4,300 hours (or 97% of the potential) of sun a year. Africa, with 60 million terawatt hours per year, is home to 40% of the global solar power potential.

Is the use of heavy metals in solar panels necessary?

使用重金属制造太阳能电池板是必要的吗?

Can we have a new technology, without knowing what to do with it at the end of its use?

我们能够在还不知道一项新技术在使用期结束后如何处理的情况下，去使用这项技术吗?

Should an older person still pursue dreams?

年纪大一些的人还可以继续追求梦想吗?

Will there always be a new and better technology?

总会有更新且更好的技术吗?

Ask people if they like using solar panels. Ask people if they know that most solar panels contain a lot of heavy metals. Also find out if people who have purchased panels, know where and how to dispose of them at the end of their useful life. Draw up a list of the many benefits of solar energy. Next, look into the drawbacks. Now share your insights with friends and family members. Make suggestions about what needs to be done to avoid the negative, so you can all focus on the positive.

询问人们是否愿意使用太阳能电池板。问问人们是否知道大多数太阳能电池板含有大量重金属。此外，还应了解购买了面板的人是否知道在其使用寿命结束时处置它们的地点和方式。列出太阳能发电的好处与缺点，与朋友和家人分享你的见解。针对为了避免消极影响而需要做的事情提出建议，这样你就可以把注意力集中在积极的方面。

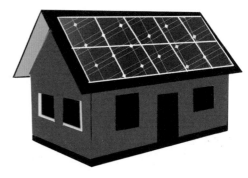

学科知识
Academic Knowledge

生物学	把藻类卷入薄膜制成太阳能电池板；太阳是地球上生物过程和化学过程的中心。
化 学	太阳核心的能量是氢原子（H）在极高的压力和温度下转化成氦（He）原子核；四个质子（氢原子核）结合为由两个质子和两个中子构成的α粒子，即氦核；镉的使用。
物 理	肖克利－奎伊瑟极限用于描述太阳能电池能量转换极限；太阳通过大规模聚变反应产生能量；太阳能电池板产生直流电。
工程学	多晶硅和单晶硅；非晶硅；有机金属化合物；薄膜太阳能电池；半导体制造厂生产太阳能电池；防反射涂层；太阳能跟踪器；峰值负载和基本负载电力之间的差异。
经济学	相比规模经济，小批量创新的制造成本更高，即使这些创新效率更高，而且能量源自可再生能源；太阳能电池板的多重好处：用作结构屋顶和绝缘层的同时能够发电。
伦理学	做错事和疏于做好事之间的区别；生命是一条道路，而不是一个最终目的地；引入可再生能源，不仅是为了节约化石燃料，而且是为了提供以前没有的救生服务。
历 史	1839年，法国物理学家埃德蒙·贝克勒尔发明了光伏电池；1881年，第一块商业太阳能电池板投放市场；1954年，第一块太阳能硅电池诞生。
地 理	太阳周围的宜居地带；光球层，肉眼可见的太阳表面，带负电荷的氢离子数量下降。
数 学	太阳发射的能源估计为384.6尧它瓦（3.846×10^{26}瓦）；世界能源消耗公式为：N为世界人口，GDP/N为世界人均GDP，E/GDP为世界能源强度（单位GDP能耗率）。
生活方式	以可再生能源满足能源需求。
社会学	在一个氏族、部落或社区中英雄担任的角色。
心理学	让人们知道你为他们感到骄傲的重要性；表示希望把改变灌输给别人；害怕失去欣赏和认同感；随着时间的推移，定下并追求一个目标的困难度也有所变化。
系统论	几乎没有合适的回收系统，导致在未来几十年会产生大量废物。

情感智慧
Emotional Intelligence

孙 女

孙女向爷爷表示了自己的钦佩之情。她敢于向爷爷提出疑问。她以坚定的方式明确表示希望太阳能发电可以变得更好。她正在敦促爷爷以更深层的方式思考，当爷爷担心自己年龄太大时，她鼓励他说，现在就心安理得地满足于自己已有的成就还太早了点。她愿意和爷爷一起追求梦想。为了让爷爷开启新的挑战之路，她开始提问。她一步一步地引导爷爷考虑太阳能电池板的回收安置问题，当她明确表示爷爷将成为自己的英雄时，爷爷也得到了回报。

爷 爷

爷爷觉得自己是应用太阳能电池板的先驱者。但他孙女的观点使他内心有点担忧，因为他想不到自己做错了什么。通过提醒孙女，自己是第一个开始使用太阳能的人，爷爷想表明自己曾在社区中拥有良好的影响力。爷爷承认自己的精力已经到了极限，不再热衷于追求挑战了。当孙女说满足已有成就还太早的时候，他想知道自己能否接受新的挑战。当他的孙女提出一个新的概念时，他很想做些什么来继续做她的英雄。他很看重孙女的爱和欣赏。

艺术
The Arts

有什么能让我们在艺术项目中使用旧太阳能电池板的点子吗？我们能用这些为发电而造的面板来创造美。在成年人的监督下工作，因为这些面板可能含有一些重金属，对面板进行操作时要小心。从材料本身寻找灵感，创作出一些闪闪发光的艺术品吧。

思维拓展
Systems: Making the Connections

生命中的一切都会结束。在构思新产品时，我们应该把生命周期纳入考虑。大自然是慷慨的，当地可找到所需的所有材料。自然界的另一个特点是，没有东西会完全消亡，也没有东西会被凭空创造出来，万事万物都会不断转化。不幸的是，我们只使用了原材料的一小部分，在其寿命结束时将余下部分视为废物处理。新产品出现时，即使是像太阳能电池板这样的能源转型必需品，我们也应从最开始就想好产品使用后该如何处理。轻率地丢弃含重金属的产品会对人体健康和生态系统造成严重破坏。

人们应该设想这些产品在报废后如何实现价值转化而非单纯回收。在设计产品时，应认真考虑如何将这些材料（无论有毒性或无毒性）在寿命结束时转化为有用和有价值的产品。如果我们不这样做，那么我们不仅有可能污染我们的身体和环境，而且会失去材料中隐藏的价值。这会导致效率下降，导致我们的商业、社区和社会恢复力的下降。

动手能力
Capacity to Implement

你如何让人们相信太阳能是他们需要的解决方案？让我们研究一下劳拉·斯塔切尔的情况。她不从事太阳能领域的工作，而是从事医疗保健工作，她在工作中通过对太阳能的利用降低了婴儿死亡率。这是一个非常实用的方法。所以，像斯塔切尔博士一样，找出非洲和其他发展中国家尚未解决的两到三个主要问题。确定如何利用太阳能改变人们的生活。应当从一开始就寻求可再生能源解决方案，而不是仅仅用可再生能源替代煤炭等燃料。这种基于解决方案的方法将为你提供更好的改变世界的机会。

故事灵感来自
This Fable Is Inspired by

劳拉·斯塔切尔
Laura Stachel

　　劳拉·斯塔切尔在欧柏林学院获得了本科学位，并于1985年在美国加利福尼亚大学旧金山分校获得了医学学位。此后她从加利福尼亚大学伯克利分校获得公共卫生硕士学位。她在尼日利亚工作时，注意到那里由于缺乏电力，孕产妇的死亡率很高。她专门为当地一家医院的产房、待产室、手术室和实验室设计了一套太阳能发电系统。这使得第二年的孕产妇死亡率下降了70%。斯塔切尔博士建立了非盈利组织We Care Solar（正式名称为WE CARE，"女性紧急通信和可靠电力"的缩写），提供了数千个太阳能手提箱设备，服务了数百万名母亲。

图书在版编目（CIP）数据

冈特生态童书.第七辑：全36册：汉英对照 /
（比）冈特·鲍利著；（哥伦）凯瑟琳娜·巴赫绘；
何家振等译. —上海：上海远东出版社，2020
ISBN 978-7-5476-1671-0

Ⅰ.①冈… Ⅱ.①冈… ②凯… ③何… Ⅲ.①生态
环境–环境保护–儿童读物—汉英 Ⅳ.①X171.1-49

中国版本图书馆CIP数据核字（2020）第236911号

策　　划	张　蓉
责任编辑	程云琦
助理编辑	刘思敏
封面设计	魏　来　李　廉

冈特生态童书
太阳能利用的未来
[比]冈特·鲍利　著
[哥伦]凯瑟琳娜·巴赫　绘

贾龙智子　译

记得要和身边的小朋友分享环保知识哦！
八喜冰淇淋祝你成为环保小使者！